Authors - How To Brand You and Your Books

By

Jolene MacFadden

Paperback ISBN: 979-8-9867691-1-0
Hardback ISBN: 979-8-9867691-0-3
eBook ISBN: 979-8-9867691-2-7
Audio ISBN: 979-8-9867691-3-4
Library of Congress Control No:

Southern Dragon Publishing
c/o Jolene MacFadden-Kowalchuk
PO Box 1712
Mayo, FL 32066
https://southerndragonpublishing.com

Printed in the United States of America

First Edition

TABLE OF CONTENTS

A Note For the Readers

This book started out as a booklet. It is still available for purchase called 'Ten Ways to Market Your Book For Free'.

I have been gaining more knowledge about the different digital marketing techniques for myself and my clients. I want to assist each of my fellow authors who are published or not, to learn more about 'branding themselves and their books'.

There is much to learn about becoming a published author and taking command of your own book business. There are a lot of free resources available as well as some expensive ones. I hope to provide my readers and all my fellow writers enough information to help you get started.

In the back of the book there will be some links to my website which contain a wealth of FREE Bonus information, digital and print forms, and other items to help you along your book business branding journey.

Thanks for purchasing my book and please sign-up for my newsletter to keep up with all the great ways we have discovered to gain more followers and purchasers of our stories.

Introduction

Hello everyone. My name is Jolene MacFadden. I am a blogger, website designer, social media coordinator, published author and owner of several websites. This includes the North Florida Writers Tour where I feature published writers who live and/or work in the North Florida area. The tour includes fun and informative articles and promo videos that authors can use in their own book marketing strategy.

If you find this book helpful, I would love to hear from you. If you would please start by writing a brief review on the bookstore page from which you bought this book and add that to Goodreads.

If possible, create a video on your TikTok, Instagram and/or Facebook pages. (A great way to start your own digital marketing is by supporting your fellow authors.)

I always appreciate those who join my newsletter list on my website(s) and subscribe to my other social media pages. I share my own reviews of books I have read, write up tips and strategies in digital marketing and creating your own brand as well as what is happening with my own writing journey.

My Writing Website

https://jolenesbooksandmore.com

My Publishing Website

https://southerndragonpublishing.com

Free Book Promotions and Articles for North Florida Writers

https://NorthFloridaWritersTour.com

Jolene's Writing Twitter Page

https://twitter.com/JoleneMacFadden

Jolene's Author and Bookstore On Facebook

https://www.facebook.com/jolenesbooksandmore

Jolene's Author and Bookstore on Instagram

https://www.instagram.com/jolenesbooksandmore/

Jolene MacFadden's Profession LinkedIn Profile

https://www.linkedin.com/in/jolene-macfadden-kowalchuk-29a55825/

Jolene's Writers YouTube Channel

https://www.youtube.com/c/jolenesbooksandmore2669/

Jolene's Online Bookstore:

https://bookstore.jolenesbooksandmore.com/

Join in on the Fun at Jolene's Patreon:

https://www.patreon.com/JolenesBooksAndMore

I am even creating a Jolene's Writers and Book Talk Podcast

https://podcasts.apple.com/us/podcast/jolenes-book-and-writer-talk/id1619792506

Remember we are all trying to figure out how to create and grow our book business in an ever-changing market. Subscribing, Liking, Sharing and Commenting on each other's social media pages and blog/website articles helps us all move forward together.

What the Heck Is Branding Anyway?

In its simplest terms, Branding is a marketing term: 'Brand, a name, logo, slogan, and/or design scheme associated with a product or service' This comes from WIKI.

It is the same on any online dictionary that you search. Put another way, '*Branding is the process of creating a strong, positive perception of a company, its products or services in the customer's mind by combining such elements as logo, design, mission statement, and a consistent theme throughout all marketing communications*'.

Everyone is worried about setting up their 'Brand' or being consistent with their 'Branding' when they create social media pages and websites. Since your writing is your business, it really is up to you to choose the color schemes, font styles, and any logos you create so that each is reflective the image of yourself and your books you wish to present to the world.

You can use the font styles and some colors in every post that you create. You will have the added burden of trying to stick closely to the styles associated with the genres in which you publish as well.

Keep in mind that people of all ages will need to be able to read any posts and messages. Try to stick to fonts that are not too artistic and colors that more people can distinguish. Check out the following:

https://www.webmd.com/add-adhd/features/what-is-neurodiversity

https://elementor.com/blog/website-color-schemes

/

Book Information Pages

As part of your first marketing step, you will need to keep handy certain information about each of your books. One straightforward way to do that is to create an information sheet with the following information.

Book Title: _____

Pre-and/or-Publishing Date: _____

Book Formats Published: eBook / Paperback / Hardback / audio

Book Purchase Links_____

_____

_____

_____

_____

Synopsis: _____

_____

_____

_____

_____

List of the Main Characters   Minor Important Characters

_____   _____

_____   _____

_____   _____

_____   Genre/Tropes_____

Describe the setting of the Book

_____

_____

_____

_____

_____

30 Word Elevator Pitch: _____

_____

Another suggestion would be to have a folder on your cloud or hard drive entitled with your books name or simply 'My Books' and create Notepad TXT document of all the information, website and social media links and any other pertinent information for each book.

The reason for the text document (book1.txt) is that it does not hold any formatting only text. Word and PDF documents contain coding in the background of the file that gets attached to your letters

Sometimes the other types of documents have formatting codes attached to the words or paragraph that will interfere with a clean pasting of that link or those words into another website. This is true with different versions of your word processing programs as well. The program needs to have clean text in order to help your words to look right on that website and your link to work properly.

Nasty coding people in the past figured out how to add viruses to these documents and encode other things inside the links directing people to the wrong website. If you would like to know more about how to get a clean copy of the buying link from most websites, please send me a message on my website: https://jolenesbooksandmore.com/contact-us/ and we can set up a quick Zoom of Facebook Conference where I can show you some examples.

Beginners Notes

Every single writer I know, would rather do the most hated chores than try to figure out a good strategy for branding his or her books. Then we all are supposed to implement that plan.

I don't know about you, but I would much rather sit in my office every single day typing away on my laptop authoring the next great novel. I like to create my own worlds, characters to interact in those worlds the way I want. I would even endure 100 pages of blistering comments from the editor than have to do MARKETING!

However, if you ever want to make a sale you must tell people about your book. And that my friend, entails you branding yourself, your book, and your series in such a way that is memorable and inviting to readers of your genre.

Even authors with a publishing house contract will have to do most of their own marketing as well as what the publishing house sets up. This usually includes attending book-signing events, taking part in interviews with the local press, talking to book groups, and letting their readers know that there is a new book in town.

Self-published writers will have to do everything themselves or pay someone to do it for them. The upside to the extra work involved is that each writer has complete control over the image he or she wishes to present.

There are ways of getting the word out that your story has been published online that do not cost anything. Some of these things will be done for you by whoever publishes your book.

Amazon, Smashwords, Google Play, Ingram Sparks, Barnes and Noble, or even iBooks will present your book to the public, but they will not include it with their roving sales pitches after the first 30 to 60 days without you paying them for ads. You could always sweet-talk all your friends and family into writing reviews on these sites. Popularity begets popularity in the book business.

Some Extra Points

We often learn most things through trial and error. We look for advice from successful booksellers and writers. Sometimes, we just plain keep trying things until a formula is finally found that creates the desired result.

I suggest you keep a diary of what you have tried and your success or failure with each approach. Keep track of where you applied your strategy, the results, and how long you waited for results. Everything takes time but most of what we do will either have an impact or it will not within a week, two at the most.

It would be nice if we could just announce that our book is published on this or that retail venue then all our friends and family quickly buy a copy, read it right away, then write a wonderful review. They will, of course, share it with all their friends, and they with theirs, etc., etc., etc.

But that rarely happens in the real world. It is what we aspire to and might eventually get once we have instituted some of the tips below and determined what works for our style of writing.

One thing I didn't include in the presentation that might be of interest here is the ability to create, join, and take part in public or even private groups on Facebook. A group page can be created by any person with a personal profile on Facebook and the page can be used for any purpose. You can make it private and by invitation only so that everyone must be approved to join or public so that all who click to join are automatically approved. Either way, only those who have joined will receive automatic updates on new posts.

For instance, say you are a Science Fiction/Fantasy writer, and you are creating a whole new world with fascinating characters that go on adventures. Eventually you will need beta readers and, hopefully will gain hard-core fans. Beta Readers can help you make your story rich and engaging. They are in addition to the various editors that you might employ to help you. Beta Readers can give you a unique perspective into the world you created; the characters within it; and your storyline. Plus, they are FREE!

One way to encourage people to become your Beta Readers and possibly even your ARC Readers is to create a group page for that book or series. You name it something appropriate, create rules about participation and posting, upload some of your artwork, and ask for your friends and family to join the group and start making comments.

You should also gain ARC (Advanced Reader Copy) readers. Those great individuals who will read your stories prior to publication and write reviews during the pre-release period to help build momentum prior to publication. You can obtain these volunteers in your reader's group or on other social media pages you have created as well as your website.

Another tip about Facebook Groups is you might want to join a few yourself. There are groups for everyone, every hobby, every genre, every location such as country, region, state, county, and city out there. People love to gather and talk about their favorite things. Some just want to complain but most would like to share, get to know each other, learn new stuff, and help others along their journey.

Go to the Facebook search box and type in a subject you are interested in and the word 'group' and see what comes up. You may want to join small intimate writing groups or large book advertising groups, or helpful genre groups that allow you to advertise your published works too.

The more people in the groups and the more posts per day means they are extremely popular or just too bloated with non-sense to be of much help. You must read them to see if they have regular postings, lots of comments, and interactions between members. Those are probably going to be the most helpful.

Remember groups on Facebook have rules and require you to join. They can be public or private. Always be aware of who you are interacting with and never give away personal, or financial information on these sites.

That is one of the main reasons why I have so many email addresses. One for each website and sometimes one for each function I want to present to the public.

Examples:

webmistress@domain.com for tech support, customerservice@domain.com for sales help, and Jolene.macfadden@jolenesbooksandmore.com for me as an author contact.

Remember you are going to get out only as much as you put in. If you just lurk on the edges, reading posts, liking some, now and then, and never commenting or posting yourself, you are not taking part. You must jump in and give some advice, ask some questions and, please, remember that not every interaction will be positive.

DO NOT INTERACT with the crazy ones. If someone is bad-mouthing something or someone then just walk away and move on. Sometimes the inmates do run the asylum.

Writers Goals for Branding

Ideally, we would like to keep our readers loyal from the time they discover us and our stories and for generations to come. A good marketing strategy will help ensure that today's readers will share their love of our stories with their family, friends, and to their future generations. Before we get to the practical details of the subject just a note here to emphasize a few things.

Not everything you try will work.

You should try lots of things to see what will work. Just hoping and praying will not do it.

You DO NOT have to try every single suggestion to find something that works for you.

It will probably take a combination of different things being done consistently to find just the right way to market your books to your readers.

You will find your niche if you just keep trying.

It never hurts to gather a dedicated group of fellow authors and start your own marketing strategy consortium. You ALL will need to like, share, comment, read each other's books, write reviews, and generally cheer each other on!

You are NEVER in competition with other writers you should be helping each other as much as possible.

Readers read many different genres, some, like me, read every single day. I read everything from fast articles to long books about romance, mysteries, biographies, science fiction, and even a western or two. When I'm not reading, I'm writing. Heck, sometimes I am doing both at the same time.

First Piece of Advice

Every single author out there really should have a web presence. Even if you just have a simple free website that tells the readers about you, your books, what you are currently working on, where you will be in the future to sign books and where they can buy your stories today. Yes, some publishers will create a page for their currently publishing writers, but those pages can be taken down if you haven't published in a while. You need a place of your very own to gather loyal fans.

I have listed four companies that I have used that still offer a FREE version of website creation and hosting services below. For those just starting this might be an effective way to go. Each of them has its limitations, complications, and even joys to work with. You will need to do your research on which one works best for you. Be aware there may be more but read their requirements to make sure they fit your needs, their reviews to ensure their integrity and that they have been around longer than a year or two.

This is by no means a complete list. One or more of them may not offer this service in the future. Again, do your own research if you must start with a free website. Each will have help pages to aid you in creating and personalizing your website. You will not get your very own domain (such as www.myauthor.com) for free, as your web address will have some part of their company name inside it (such as www.myauthor.freewebs.com). You will be able to keep it for as long as they offer that service.

Freeserver and Webs.com

I had a couple of free websites to start with that I mostly used for family updates. One is still there and running even though I can no longer update it. I ran into a size limitation and didn't want to delete anything, so I just left it there from 2004:

http://www.jolenespage.freeservers.com/

I had one with WEBS.COM for years but it has since been removed because they changed the rules about hosting free sites. Since it was for family only with posts and pictures for the years 2004 to 2009, I just didn't think to keep checking it. Both companies create web pages.

From Webs.com

How long will my site stay free?

As long as you like! Your site will remain live for free if you continue to login to your Web account (I recommend logging in at least once a month). *Some users stick with free for life, and some choose to upgrade to Premium as their needs grow.*

For those who may not know, a blog is a type of internet page where you can post daily, weekly, or monthly articles (updates, news, etc.) and they will appear in the order you post them with the newest on top. You can add 'static' pages (those that generally do not change much) to the blog but it functions mainly as a digital journal.

A plain website consists of one or more static pages with information you can publish and never change unless you desire to update them. Some programs/hosting services allow you to add the function of the blog or journal to that set of pages, but it will not be the first thing a visitor sees.

Blogger

Since I have had a Gmail account for several years, I also created a family blog site using my personal Gmail account. It was easy to learn how to work but, remember, use those help pages and read through those online forums to get your blog site to look the way you want it to look.

There is some coding knowledge required for some functions. The FREE theme selections are limited. I would get bored with how it looked and change the theme every couple of months.

Remember this is a BLOG and not a website. You can add web pages to the blog but posting regularly is the function of this space. You will have to create or currently have a Gmail account to use this service. The good news is that is FREE as well. https://www.blogger.com

Update: Google also supplies a free service called 'Sites' It can function like a website.

Wix

The 'latest and greatest' website creating software and hosting company called WIX.COM has FREE website creations and blog capabilities too. You see it on commercials every day and talked about on most of the social media platforms. They make it look easy on television, but I did not find that always the case. They have a strict layout choice for the sites. The themes and color selections are limited and some of what they offer just really does not look right on every device. They are, however, very smartphone friendly.

I created a couple of different free sites for some writer friends just to see how it would work. It took several hours and a great deal of searching on the internet for more help to get it to look the way we wanted. Then once it did, we couldn't easily change it to something else. So, I would say it has a very steep learning curve but is doable if you don't mind the limitations.

Wordpress

WordPress is considered a 'Content Management System' software package. It was created as freeware and must be hosted on a web server somewhere to work. It can be loaded to almost any kind of Webhosting service (most of which you do have to pay for). There are two different websites with the name WordPress.

The first is the original creator community at https://www.wordpress.org They keep the software updated; house the free plugins you can add different functions to your site that you might need as well as 1000s of different free themes. They have hundreds and hundreds of 'Help' pages and forums to ask and answer any questions you might have.

The commercial website at https://www.wordpress.com host the websites for the community. They offer a basic FREE website hosting version, but it is fully functional, has access to all the free plugins and themes, and is already setup with an SSL certificate.

SSL stands for 'Secure Socket Layer' is an added security layer built into the website address. It makes your website address change from 'http' to 'https' and is now required by most search engines. Every hosting service now should offer that service for free.

Wordpress.com also owns and administers one of the most popular plug-ins in the world called Jetpack. It has a lot of free features. The paid ones offer to improve security and your website functionality. The free version allows you to automatically shares posts and pages to the main social media services currently supported such as Facebook Twitter, LinkedIn and Instagram.

For instance: When you post a blog post on your website after you hit publish the Jetpack program can automatically share that with your Facebook and Twitter pages. This will, hopefully, help you increase your followers more quickly as your audience grows from the various techniques we will discuss later. The free hosted site does have its size limits as do all of them. Also, should your visitors exceed a certain quantity every month you will have to upgrade to one of the paid services.

The one good thing about starting with Wordpress.com is that should you ever want to upgrade to the paid hosted account with your very own domain they will do all the work for you when you switch. You will not have to download a backup and reload it back up to the new domain. [Although you should do that every month anyway.] They have conversion software that walks you through those steps. Should you wish to transfer to a new hosting service you can export your website to your desktop and import it to the new service once the domain has been bought and the WordPress CMS loaded.

That, my dear readers, is a whole other conversation and I charge for those. Please see my Southern Dragon Publishing (https://southerndragonpublishing.com) website for a list of the services I provide along with the prices for each. If you sign up for my newsletter, I will, of course, post discount coupons occasionally.

Google Sites

A new comer to the field that is another project by Google is called Google Sites. It is a different division or application from Blogger in that it creates a website to showcase you and your books as a business or portfolio. You will need to create a Google account. It works the same as Blogger if you have a Gmail account. The creation process uses a drag and drop process with a slim set of themes to choose from. If you get brave you can experiment and add some of the more complicated features but for just adding text, pictures, and clickable button links it is fairly simple to begin.

Check out: https://sites.google.com

You can start and finish within a few hours and once you have 'Published' your site you can add the link to your social media pages, groups you participate with and wherever you need to add a place for people to find you. After doing a quick review the setup is simple and makes use of the same kind of structure as Google Docs, Sheets, and some of their other products.

Remember that Google own these pages and hosts your site. You will want to keep copies of what you create. Whenever the company decides to cancel the program, you will lose your site. Granted, Google generally takes quite some time to phase out any of their applications so you will have plenty of notice.

SOME FINAL POINTS

If you would like assistance with setting this or any of the free websites up, we can do a Zoom conference and share screens. Yes, I do charge for my time, but I have discounts for these types of services all the time. Please check my Southern Dragon Publishing website and join my newsletter.

Once you have a website, whether it is free or not, you will want to keep it updated at least once a week especially if you have a blog of any kind. Check to make sure your information on your books is correct and you have the best-looking cover on your site, and they are linked to the page where your readers can buy copies.

Check it on your smartphone, your iPad as well as your laptop and desktop. Make sure all the links go where you want them to go. And occasionally, refresh your 'About' page as well as update the marketing materials you have created and have available for download.

My website address:

_____

Account User Name:

_____

Password:

_____

Have You Typed in Your Writer's Biography? Yes/No

Creating different versions of your biography and updating it regularly helps keep your 'About' sections for each social media platform unique and interesting. It also helps to have different lengths of the biography available inside your press kit. *(The elevator pitch 30 to 100 words, the mid-grade with about 150 to 300 words and a nice comprehensive one of about 500 to 1000 words.)*

Have you Uploaded Assorted Sizes, high-quality, professional, Author Pictures for the Press Kit? Yes/No

Getting your picture taken by a professional conveys that you are taking your writing business seriously. Take advantage of local discounts when you can. Call around to the local photographers to see if they will work a deal for a short outdoor or studio session. Ask them about makeup and clothing choices prior to scheduling. If you are good with makeup, they can suggest small touches to help improve the picture. If not, some even offer applying makeup in their studios for a small fee.

However, if you are just starting out, you can shoot the photos yourself using today's high-resolution cameras available on most smartphones. Your camera should have at least 6MP or 6 Megapixels:

*(MP), which translates to "**one million pixels**," dictate how much detail your camera's sensor can capture. The higher the megapixel number, the more potential detail can be captured within an image.*

Digital cameras on most newer devices have a minimum of 6 or higher. The higher the better. Also, indirect natural sunlight pictures, generally, come out very crisp.

Approach these self-shot photos as a professional would. These are not casual selfies. Dress professionally in neutral colors with no extra frills. Apply light make-up to flatter your face. Hang up a black cotton sheet or curtain, preferably outside on a sunny day but not directly in the sun. Have someone take at least a dozen different pictures with the highest quality settings either on your cellphone or a digital camera.

You can then take those pictures, remove the backgrounds and place your body on any type of background you wish. There is a free site that will remove the backgrounds from any photo you upload. You should not have to do any editing for the removal unless you wear a color that matches the background. https://www.remove.bg/

Another tip for your future branding journey would be to sign-up with https://www.canva.com/

They have a free version and an inexpensive professional version that allows you access to a wider variety of pictures, videos, music and more. I use it every single day. Try the free version first. It seems to be as easy to master like WORD or POWERPOINT but way easier than any of the ADOBE products. The preset templates alone are well worth the professional monthly fee.

Have You Created a Page for Each of Your Books with Pictures of the Covers and a good synopsis? Yes/No

You may have only published one book or none yet, but you should have a place with the name of the book in the link to help you begin to spread the word about your stories. Each page gets a separate rating in the search engines. This helps others find what they are looking for and bring them directly to you and your story. Make sure you have a couple of different views of your book along with a detailed synopsis. You can add some nice review quotes from reviews of the book to this page, but a good synopsis is still required. You might want to add your book trailer video and links to all of the book purchasing sites or a Universal Link Button. (more on that later.)

Have you created a Contact Page with either an online form connected to your email or just your email? Yes/No

This may seem a little redundant. Creating a separate page that is listed as 'CONTACT' is one of several pages that are required for a well-organized website. You can use an online contact form for a variety of different marketing chores including gathering your email list. You need to have a listing of those who like you and your books. They will be your first line of communication for all the new books you publish.

Have you created a Universal Buying Link (UBL) for each of your books? Yes/No

https://books2read.com/ will help you create Universal Book Links for free. You can list all the places the books are available and in every single format with just one link. It takes a bit of time, and you may find your books are being published in places you didn't know about. These links can even be personalized.

Here is one I created for my Volunteer Workamping book:

https://books2read.com/InsiderGuideCampgroundHostingFla

If you are writing a blog or emailing a newsletter or both. you can add these links in to encourage people to explore your published works in all its many different formats.

Remember you will need to create a mixture of short fun posts or longer articles at least once a week to keep your followers engaged and coming back to your website. Social media posting is another animal altogether. There is more on that later in this book.

I suggest that you write up some article suggestions that you think your readers might enjoy. There is a digital form available for you to print out on my website. See the Bonus Materials Section for more information.

Topic /Notes/Date Posted

_____
_____
_____
_____
_____
_____
_____
_____
_____
_____
_____
_____
_____
_____
_____
_____
_____
_____
_____
_____
_____
_____

Amazon Author Central

Whether you self-publish or have a publisher you should make sure that your Author profile page is complete with the latest information about you and your books that are currently available.

Amazon Author Central is FREE to every published writer.

They create a simple internet link for you to share on your website and social media pages.

You can make sure that all your published works are connected to your profile.

They list your sales reports of your self-published books

They allow you to attach your blog site to your profile so that, when you post updates and news, it will automatically be shared on your Amazon profile pages.

Again, this is FREE.

I cannot tell you how many times I have been researching a writer after finding some interesting book cover or book synopsis that I wanted to try. I get to the selling site (Amazon mostly but check in with Barnes and Noble, iBook, Google Play, etc.) and there is nothing or little about the author. I mean they have the books selling there but there is no information about them on the site.

Come on people! The most basic thing in selling is to make it easy for your current and future customers to find you.

With a website you want to make sure the pictures you have of your books are linked to their selling pages. If you are selling books on more than one service, then download their icons and attach the links to your books to those pictures. Of course, you can also just use the Universal Book Link you created with Books2Read.com.

We will assume that you have gotten a professional headshot uploaded recently along with an updated, fun, informative short biography to add to each site. Each retailer can assist you in getting each version of your books attached to your author profile.

Marketing 101: tell people what you are selling, show them who is selling it to them, and supply interesting tidbits of information so that they can relate to you as a person.

We do not need to know your dress size, your children, or grandchildren's full names, or even your partner's. You can give them their own pen names when you talk about them. Even general information such as your hobbies, what you like to read, etc. is fun for your fans to read about as it is with any celebrity, we wish to know. That makes you (as the celebrity in question) more relatable to your potential readers and fans.

I usually save this kind of information for my website design and social media coordinator clients but one piece of advice I do give to anyone:

'Keep your personal life private but let people know you are human and have experienced some of the same trials as everyone else.'

Having a sense of humor about those problems will go a long way in helping people to know you. Telling stories about your pets and children or partner is always fun. I mean just talk about something funny they did or said but not really who and where they are.

A tip that you might not be aware of is that from your Amazon Author Central page you can add every single edition of your books yourself and then ask that the paperback, eBook, and audiobooks be combined into a single listing page. You may have to add any new books you have published recently if they do not automatically appear after you publish them. And, finally, any backlisted book titles that are still out there that you haven't republished can be linked as well. You may not want to republish them right now, but you do want people to find your new books if they loved your old ones.

Everywhere your name appears on the internet you hope to have a professional-looking picture showing, a brief bio, and information about your published works. If you have a publisher, you want them to supply a professional-looking page for you. Your profile page for each of the major booksellers (if they offer it for FREE) should be completed as soon as possible and updated yearly.

Finally, having a publicity package together in PDF format for download to any newspaper or media outlet that wants to interview you or host a book-signing event is helpful as well.

Try to change the Bio for each

Amazon Author Central

Profile Link: _____

Pictures added Yes/No

Video Uploaded Yes/No

Linked Your Blog Yes/No

If you would like to see a sample, you can view my profile on Amazon. They create a short link for you, but it looks different once you paste it into the browser.

This is the short link

https://www.amazon.com/author/jolenemacfadden

This is what it gets converted to

https://www.amazon.com/Jolene-MacFadden/e/B00O2JNQ7Y

You can use either one, but you might as well put the first one because it has your name and the word author. This is great for letting people know you are an author and helps with your SEO (Search Engine Optimization) rankings.

There are not that many places that you publish your books that will create an author profile page for you. When you find one that does, please take advantage of that service on each of those selling sites. You just never know when someone will be looking for your book and want to connect with you as a writer.

Smashwords

This was a wonderful place to publish your eBooks. They have been supporting Indie Authors for a long time. They do have specific requirements for uploading your manuscript to pass their software editors. However, they provide you with a nice manual which has simple instructions for preparing the WORD.DOC prior to uploading.

They have a great Author Page section to add your picture, a biography, answer some of their FAQ questions for your fans and now you can upload videos.

They have a coupon tool and a free book buying widget that you can use on your website.

They have a section at the bottom of the book page to add links to where your customers can purchase the paperback and audio versions of your book.

Smashwords publishes and distributes your books out to a wide network of retail stores as well as the Library Network. Once approved, you will be given links directly to each of your books on the retailers buy page. Of course, if you add your Smashwords book link to your Books2Read Universal Book Links page (see below) that program will automatically add all the other retailers for you.

Here is my updated Smashwords Profile page

https://www.smashwords.com/profile/view/jolenemacinjax

Did you know that Smashwords allows you to upload videos to your profile page about each of your books? Yes, they do! And, it can be either your book trailer, you reading from your book or even a promotional sales video for one or more of your books.

Link to Smashwords Author Profile:

_____

Profile Picture Added Yes/No

Biography Added Yes/No

Video Uploaded Yes/No

Answered All Interview Questions Yes/No

Draft2Digital

I will say that Smashwords may be going away soon as Draft2Digital has bought the company. The last update from both companies said nothing will change for a little while yet. You should go ahead and create your author profile on Smashwords too as it gives you access to more marketing tools.

It would be prudent to check out their FAQ page for updated information *https://draft2digital.com/faq/*

Draft2Digital creates an author page for each of the books you have published with them on their Books2Read platform. You are encouraged to spruce it up on Books2Read.com with a profile picture, a biography and your social media links. Then with each book of yours that you create Universal Book Links for, no matter when or where they are published, they will be connected to that author profile page.

I have filled out my own. And if you click on the link below you will see that each of my books I have published so far is connected to my profile. Click on the Buy Now button on any of them and you will be taken to the Universal Book Link Page for that book with every single link currently available for purchase. That is a nifty marketing feature.

https://books2read.com/ap/x21KL1/Jolene-MacFadden

You are able to use your Draft2Digital Login with Books2Read that way any books you publish with them will automatically update on your profile page.

Books2Read Author Profile

Link to Profile: _____

Profile Picture Added Yes/No

Biography Added Yes/No

Links to all your Social Media? Yes/No

Facebook Pages

If you do not want to join any social media pages you do not have to, however currently, it is especially important that you take the time to at least create a Public Author Facebook Page. This is separate from your personal profile account with Facebook that you must create initially. The personal profile is strictly for you, your family and personal friends to share news, pictures, and video. Make sure you turn off public sharing in the Privacy Setting of your Personal Profile page.

Public business pages do not require people to become 'friends' with you to receive updates. Your readers only must choose to 'follow' and/or 'like' the page to start receiving updates. The Facebook Public Page is just that, public. Facebook will give you a public link to share on your website. You can also add a link to your Amazon Profile page, and any other internet pages that have your profile attached.

Creating the page is simple and there are instructions as well as videos available to help walk you through the process. I would suggest that you choose the Business/Brand page. Your writing is a business, and you need to treat it like one. You do not have to list your home address or home phone number. A business address (PO Boxes work too) and some other phone number would be preferable.

You can get an internet phone number through Google Voice for free. A business address can be rented through one of those postal services companies. Do add your website address, fill out your 'About' page, and read each of the sections to find out their purpose. You can remove the ones you will not use. The more you have filled out the better your page will look, and the more people will be drawn to it.

Have a nice Facebook Page header photo. It should be the same header (or similar look and feel) as the one you are using for your website and other social media pages. Upload your professional picture as well. You are trying to create a consistent image of yourself as a writer and your writing business.

Creating or having a business logo created for your websites and social media pages is also a great idea. It is not necessary to start but if you create your own publishing company to publish your books in the future you will need one.

You must have this page created before you can add it to the social media sharing function on your WordPress site. Most of the other free websites only allow you to post the link to your Facebook page. So, don't forget to go back and add it when you are done.

There is some debate of whether you should create a Public Facebook Page for each series you create. That might detract from your author profile page. Some authors have created private Facebook Groups for their series followers. It does not hurt to try that. It could be quite fun, but you will also want to get your group subscribers to sign up for your email newsletter for exclusive discounts and updates on the series.

Facebook Page as a writer and/or as a series link:

_____

Added a Professional Photo/Appropriate Avatar Yes/No

Created a Facebook header *(you can use a simple program like Paint to create one)* Yes/No

Fill out all the pages and sections as fully as possible and turn off the ones you will not use.

Ideas for posts: 2 or more times per week and one of those can be a link to your latest blog posts

Examples:

create memes (a humorous image, video, piece of text, quotes from your book on an appropriate background.),

show pictures of your pets,

take quick informative videos of your hobbies

show book cover ideas, and get feedback, etc.)

There are a variety of things you can do on your Facebook page that will help you gain readers. The best advice that anyone can tell you is to figure out a plan of action of what you are posting, when you post it and what you are going to do to add 'Value' for those that follow you. The key is to be consistent but not posting a large quantity of stuff just willy-nilly. Try to be dependable for at least 6 months before trying something new.

FACEBOOK PAGE POSTING IDEAS

YOUR FACEBOOK WRITERS PAGE:
_____

YOUR FACEBOOK GROUP
_____

FACEBOOK GROUPS YOU HAVE JOINED:
Print the name of the groups you have joined. Keep track of when you last interacted with the posts, created your own post, etc. This will help you get in a routine. I think you will notice an increase of followers on your own public author page and website with these consistent helpful interactions.

_____
_____
_____
_____
_____
_____
_____
_____
_____
_____
_____
_____
_____
_____
_____
_____
_____
_____
_____
_____
_____

Other Social Media Pages

Joining other social media pages is strictly up to you and your preferences. Each of them has their own appeal for both writers and readers. Depending on your age, your genre, your abilities and even your sensibilities one or more of these may be just what you need to gain more readers.

Each of the following are also free to join and use. Facebook owns Instagram and will allow you to connect your feed/posts from Instagram to your public Facebook page. Each picture or video should have at least a couple of sentences in the text box if for no other reason than to share something funny or explain what the picture is about. Preferably, you could start a conversation by asking one question per post.

A Quick Word About Hashtags

You will need to learn how to use (#) hashtags for all social media posts no matter which service you are using. Trust me on this. You can see which hashtags are being used by your fellow readers and copy some of those. You can even make up a few for you and your books. I do recommend you have a listing of the hashtags you like and the ones you have created for yourself in one of those handy text documents. You will not be using ALL of them every single time. You will need to make them relevant to the post you have created.

Top 10 Most Used Hashtags for Writers

#WritersLife	#WriterWednesday
#WordCount	#FollowFriday
#WritersLift	#WritingPrompt.
#BookGiveaway	#AmWriting

Twitter

Twitter is an excellent place to find other authors and readers too. Yes, you only share short bursts of information, a picture, a short video or links to your current blog post and videos on YouTube. This is where the hashtag knowledge really pays off. You can use the ones you see other authors use but you will need to create a list of your own. If you created your hashtags listing in a text document as suggested in the prior section, you can just copy and paste the ones that pertain to your particular post.

You must stay active in this community by posting fresh tweets yourself as well as retweeting and retweeting with comments. I make sure I have 'retweeted with comments', shared with comments, and loved/liked at least half a dozen other posts while I am there. Mostly, you just want to create short, informative tweets with a picture and/or a link.

You can ask a question to get more engagement. It has been suggested that you create a list of your most active followers (@nameofperson) and create a tweet with 5 to 10 of these with just a 'Have a great Day' or anything similar. This is a great way to promote some of your fellow writers, gain new followers from those that follow them and get more exposure to your own account. Plan on doing this regularly.

Don't forget to use those hashtags - Twitter started it first

From the Twitter Help Pages:

People use the hashtag symbol (#) before a relevant keyword or phrase in their Tweet to categorize those Tweets and help them show more easily in Twitter search.

Clicking or tapping on a hash-tagged word in any message shows you other Tweets that include that hashtag.

Hashtags can be included anywhere in a Tweet.

Hash-tagged words that become very popular are often trending topics.

Twitter recommends only using 2 hashtags per post though.

Tumblr

Tumblr is included as a social media service. Really it is a multipurpose social media application that you can use as a blog/website. You can use their software for free to post great pictures and short videos. There is a steep learning curve for website creating and updating.

As with all free versions of website hosting it has its limits in the size of files, number of videos it will host (that slows down their servers), and customizations.

The users are mostly artists of one kind or another such as: illustrators, photographers, painters, musicians, videographers and, yes, even writers. The authors usually post memes of quotes from their books as well as book trailers or vignettes from their stories. If you are connected with a videographer, then you can create some great action videos for this and your other social media accounts. Tumblr is an extremely visual website similar to Pinterest.

It also requires the use of hashtags.

From some of the tech news feeds I found that Tumblr was owned by Yahoo!, but in 2019 was acquired by Automatic which owns Wordpress.com. Tumblr has floundered over the last 10 years, but its popularity has increased exponentially since the pandemic started, especially with the Gen-Z-ers (those under 25). It has always been known as a rebel platform. If you want to reach the younger generation, or are in this younger generation, then starting out with Tumblr may be a great way to reach more readers.

To that end, I went ahead and created another website/blog/social media page on this platform to see if anything has changed or improved. So far it is still as buggy as it was before. It will take some patience and reading through the help pages online to get your Tumblr website/social media page up and running. But once you do, you will be able to post updates from your phone as well as your laptop or desktop computer. You can even schedule posts way in advance. There is a random post queue feature that you can turn on. It seems to be still in Beta. Tumblr will pick and post your schedule items 2 times a day until it runs out of saved posts. They recommend, at least, that many posts to keep you relevant and in the main stream.

Like most of the other social media platforms they depend on you to add relevant hashtags to sort your content. They also keep track of those you follow, comment, and like along the way. Your social stream will be created using these items.

In the beginning it will just throw some things up there until you have been on the platform for a little while. You can, of course, spend some time searching for other authors, genre specific posts, etc. Start viewing their content and hit the 'heart' button and every now and then make a comment on some posts.

As with Twitter and Pinterest you do not have to post long, lengthy articles but upload short videos, links to places and articles you recommend, photos you would like to share, quotes from your books or favorite authors, an audio file or even hit the 'Chat' button. Now that is a new feature since I used this platform, and I am not sure how that works. Do take a few minutes to type in a description for each post. Before you schedule or post the item you created make sure you have at least 3 to 5 relevant hashtags in the hashtag section of the post.

As with all the other platforms, it will take time, patience and persistence to build up a following. *As a side note:* you can connect your Twitter account to your Tumblr blog posts so when you publish, they are automatically shared. You can share your WordPress Blog posts to your Tumblr page. Be aware that if you have your blog posts also shared to your Twitter account and your Tumblr then there will be a double posting. One from your website and one from Tumblr. I am not sure how to fix that or work around it. You may find the information in the help section. Or I just might do a post about it on my own website.

You can and should create pages to share the links to your other social media accounts, set up a biography page, and another one for where you can purchase all your books. If you need help with this part, again, try the help section. There are very few FREE themes to choose from and not all of them will allow you to add a page. Be sure and preview each one before you choose.

You may be thinking that this would be a great substitute for your own website. That is a yes and a no kind of thing. In the beginning when you are trying to build a following, you can just have this Tumblr blog. Especially if you remember to create those pages I suggested in the previous paragraph. Tumblr does not have a newsletter feature and only collects the accounts of those who follow you on that platform. As you grow, you will want to pay for a domain of your own and a webhosting service. Then you can encourage all of your Tumblr followers to join you on your website. Check my services page when you are ready to get your own site.

Below is the link to my own social media/website Tumblr page:

https://jolenes-book-journey.tumblr.com/

LinkedIn

LinkedIn has a free side, and you absolutely should fill out the entire profile page. It is an online resume site where you share your professional experience, (who you worked for and what you did), you can create projects that you are working on and list all of your credentials, awards and diplomas.

You can and should add pictures, create posts about what you are working on, or just share the blog posts you have already created. More and more people, including publishers and readers, are looking at the professional writing profiles on LinkedIn to published their next book or to find their next great read. Who knows you might get your next book published or find some extra paid writing assignment from sources on LinkedIn?

LinkedIn has company profiles and groups that you can like and follow. Whenever they post an update, you will get a notice. It is a very good way to gather a circle of author friends, keep abreast of industry changes and share your experiences with newer writers. Again, it is about the connections. Keep your profile updated on a regular basis, check your messages occasionally, and connect with those you wish to interact with in the future.

From your profile page you are also given the ability to create a custom link to that profile on the site. This is very handy when you want to share it with others

www.linkedin.com/in/jolene-macfadden

Pinterest

Pinterest started as being only all about gorgeous pictures. Photos of places to visits, fashion, decorating ideas, things to buy and FOOD! There was and continues to be lots and lots of pictures about food.

Today it is incorporating short videos. Pinterest encourages people to post these short, fun, informative videos of places to visit, things to buy, how to create crafts, and preparing food to eat. Lately, there are more authors coming online to sell their books. Recently I added my own online bookstore to Pinterest in order to be able to tag the books and other things I sell when I create 'pins' in either picture or video format. This includes the copies of the books for autographs.

If you like the look of Pinterest, you can start scheduling 'pins' to post a couple of times a day as many times of the week as you can. Remember to create relevant folders or 'boards' for your posted pins to reside in.

Take the time to fill out a description for these folders/boards and only share pins that belong in that board. Followers will like your pins and hopefully comment and share, as should you, or as much as you are able. Ultimately, we all hope to get as many people as possible to 'Follow" our boards to get updates every time we post something new.

Here is my Jolene's Books and More Pinterest Page

https://www.pinterest.com/jolenemacinjax/

TikTok

TikTok is taking the social media platforms by storm. Each of the older ones are adding accommodations on their platforms to copy the style of posts that they need from their users. The 'clock app' has millions and millions of participants.

The users are mostly the younger crowd (Teens to late 20s). There is a definite trend towards the older participants (30s to 60s) who are gathering in groups based on hobbies and interests. This is a world-wide community so do not be surprised at finding authors from the United Kingdom, Australia, New Zealand and much more sharing their writing and publishing experiences.

This huge community of people enjoy reading books, sharing their writing journey, encouraging each other as well as selling their books. The romance and science fiction/fantasy genres are the largest or more prolific group of creators. However, the mystery/thrillers and historical fiction writers are growing fast. Being a lover of almost all kinds of fiction and non-fiction books I have found some like-minded souls who are all just figuring it out together.

When you create your profile page on the application don't forget to choose a professional profile picture and write a brief description of your genre. Once you have reached a certain number of followers (I believe it is around 1000) you will be able to create a link in your profile to your website page. If you choose to make your account a Professional or Business account, you will be able to add your link right away, however, you will not have access to as much fun music to add to your videos. There is a tradeoff.

Facebook and Instagram now have what are called 'Reels' and YouTube has 'Shorts'. These are short, fun videos created in the portrait size (9:16) and are anywhere from 5 to 30 seconds long. All this in a bid to keep up with the TikTok platform.

Write down your usernames, passwords, and links to each of your social media page accounts. You will be sharing those a lot.

LinkedIn.com

Your Profile Address: _____

Your User Name: _____

Your Password: _____

Twitter.com

Your Profile Address: _____

Your Twitter Handle: @_____

Your User Name or email: _____

Your Password: _____

Instagram.com

Your Profile Address: _____

Your User Name: _____

Your Password: _____

Pinterest.com

Your Profile Address: _____

Your User Name: _____

Your Password: _____

Link to a particular board you share with your readers

_____

TikTok.com

Link to Your Profile: _____

User Name/Email: _____

Password: _____

Other social media:

_____

_____

_____

_____

_____

Some Extra Tips

There are lots of different social media platforms out there to choose from as well as the ones mentioned here. In my experience, these are the ones that have been around the longest and seem to work better for writers rather than some of the others currently active. As with anything it does take time to build an audience/following and the more you put into it the more you get out of it.

We are all super busy right now. Time flies and we just forget to post on our social media pages or do not have any new material we would like to share. I am guilty of this as well but, please, do not post the same exact thing on every single one of your social media pages on the same day and time.

Here's a time saving hint.

Create a great informative video, download the transcript from that and post it to your blog. Then take snippets from the video, add some text and post it to all your social media platforms. Make each one a little different. You can also take stills from that video and put one main point or tip per picture and post every other day until you run out. Keep doing that each month and you just may have enough fresh and interesting posts to satisfy all of your social media accounts.

Goodreads is Still Valid

There are several sites out there dedicated to introducing books, old and new, to readers, sharing reviews, and creating lists of books to read, to recommend, connect to other readers and friends from around the world. Goodreads is probably one of the most popular sites dedicated to those who love to read.

If someone has already started you an author page by adding one or more of your books you can and should 'CLAIM' that page for yourself. It is a real pain sometimes to get two author pages combined and if you publish under more than one name that can get confusing.

All authors are given an author profile page where they can upload their current publicity photo, a short biography, and connect every single edition of their books for people to find and buy. They even allow the authors to connect their blog to automatically post the latest updates. This is an excellent feature for every author whether you are just starting out or have published lots of books.

Goodreads helps you to gather as many followers as you can by connecting with friends and, family in your email contact list who are already on Goodreads and inviting those who are not there to become members. You will need to write reviews of the books you have read. Yes, we need more reviews for everyone!

A Note About Reviews

As a bonus each review you write on Goodreads there are some boxes to tick **prior** to publishing it. Checking them will post your review to Twitter and create a widget that you can add to your website as a blog post. This widget brings over the book cover and the review.

Click the boxes in the bottom right-hand corner.

Click on the 'Blog the Review' box. Copy the code then paste it into a blog post. You can share your review to your Facebook page from here or from your website if it is connected.

This is what the code looks like in your blog box. Mine is set to HTML text view so I can change the look of what I see.

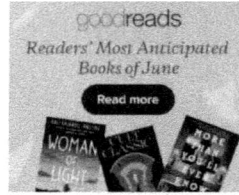

My blog is connected to my Twitter, Facebook, LinkedIn and Instagram Accounts. So, I publish the blog post about a week later to get more exposure for the book. And it is one less blog post I must create!

Below is how it will look on my blog.

Tempered Steel by Paul J Bennett

Posted On June 15, 2022

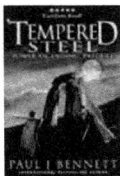

Tempered Steel by Paul J. Bennett
My rating: 4 of 5 stars

Swords, Barons and a Feudal system in some far-away land. The Power Ascending series doesn't have a lot of magic happening in this prequel but it set in the same world as the other series. I'm sure the first book will have lots of great magical battles as well as sword fights all around. This book introduces Ludwig and Charlaine. The main characters and love interests for the series. He is a Baron's son and expected to run the kingdom when his father dies and she is a daughter of a Smith. She has just earned her Masters status as a Swordmaker. Ludwig is your typical privilege Lord. That is until they meet.

View all my reviews

Finally, as you publish new books, you will want to go to your author page after about a week or so, to see if the book was already updated on your author profile page or needs to be added. When adding new books go slowly and read the entire page. If you forget to upload your cover photo you will have to ask that a librarian add them for you. Of course, you will want to link to your eBook, paperback, and then your audio versions. Once completed you can ask that the three editions be merged into one listing.

Here is a link to my Goodreads Author Page

https://www.goodreads.com/author/show/8647889.Jolene_MacFadden

Change the Bio up and include a professional photo

(don't forget to connect your blog feed)

Link to Author Profile: _____

Your User Id: _____

Your Password: _____

• Connect ALL versions of your books to that profile. You can ask the librarians for help.

• Go through and click on all the books you want to read.

• Read reviews from others and make helpful comments.

• Start writing reviews for those you have read at least once a week or more.

Book Title/Date Done

_____

_____

_____

_____

_____

_____

_____

_____

_____

_____

Bookbub and Other Listing Sites

I have been hearing about Bookbub for a long time but never have taken the time to see what they are all about. I do get a Bookbub email every day with a listing of bargain books and then later another one listing newly released ones. But I haven't looked at what they do.

This is a book bargain database site where readers can find out about new releases coming out, bargain-priced books in several different formats, and create a daily and weekly newsletter that they send out to all their subscribers. They also allow authors to create a Free profile page, connect all their books, share news about new releases, and recommendations about books the author may be currently reading.

They have paid services as do all these types of programs. I just recently got approval as an author under a different email address. I have so dang many email accounts these days. So, I am evaluating it to see just how the free service works and what is included in the paid services.

There are quite a number of different websites out there that lists books by author, include a short bio and even will post their website and social media accounts. Like me, they are book lovers but not necessarily publishers or writers. They can be useful if they have a large following for introducing your books. If they do not already have your published books and author information listed on their site, it does not cost you anything to contact the coordinator and asked to be added.

My personal favorite is **Fantastic Fiction.** I have used their listing in my research and articles for a number of years now. They keep the site updated with all the latest information; book covers being published as well as where you can buy the books. Inside specific book listings they have started adding eBay sales pages too. https://www.fantasticfiction.com/

You can join using your Facebook sign-on or your Amazon sign-on. That is convenient in that you don't have to remember another username and password. Also, it does make it easier to share any books you want to recommend or reviews you might do from there as well. The site is well laid out, not too overwhelming and has some nice graphics.

Welcome to the Fantastic Fiction

Members' Area

Now you've signed up, you can use the buttons on our author and book pages to follow authors and add books.

You can:

Follow your favorite authors (up to 100 authors)

We'll email you when we add any new books and also when books become available.

Keep a wish list of books you want to read (up to 100 books)

We'll email you when books on your wish list become available, and on your Wish list page you'll see all the books and when they are due to be published.

Keep track of the books you've read (up to 15,000 books)

Wherever you see one of your books, we'll show ✓ by the title, so you know it's one of yours. For example, on an author's page you can easily see which books you've read!

Don't want any emails?

If you'd rather not receive emails from us, please change your email preferences on the Settings page

The others you will have to find for yourself. And then you can share them on your website and social media pages!

BookBub – new bio and picture too
https://www.bookbub.com/welcome

Profile Link: _____

User Name: _____

Password: _____

Books Read/Date/Reviewed_____

_____
_____
_____
_____
_____
_____
_____
_____
_____
_____
_____
_____
_____

Share Your Autobiography on WIKI

I am an editor for Wikipedia and have been adding current information and resources to some of the articles over the last several years. One of my favorite authors is Dorothy Gilman. I created my first website for her as a fan site over 20 years ago. I have a link directly to her fan site that is still running today at the bottom of her wiki article. There is no reason you cannot do the same thing.

Most are done by fans or possibly even publishers. Create your very own Autobiography in the wiki style. They have a subcategory of Writers from each state and then Writers from the Cities within the state. It would be best that you think of this as more of a resume. And it would be even better if you can get a disinterested third party to enter these articles on wiki for you like your writer's group or association. If you are traditionally published, the company should create one for you. They may wait until you have a consistent backlist or have won a writing award of some kind.

https://en.wikipedia.org/wiki/Wikipedia:Biographies_of_living_persons

Visit the above page for more information. It is another example where other writers and the writer groups they belong to are helping each other out. It would be great to assign one person to gather the information in the form required by Wiki, learn how to post properly, and give verifiable sources. Once you have been approved and added to the proper categories you should be able to keep the page updated with every book your members publish.

Again, if one of the members wins an award or something, the organizer of the event should be able to add it to the author's wiki page. A copy of the 'official' announcement really should be sufficient as a source. You just need to reference the giver's website or link directly to the page with your name on it. The more links to other sources the better. Once added you will be able to make additions of your own every year. Once it is done it will be there for an exceedingly long time.

YouTube Can Be Fun

Now that most everyone has a smartphone with video capabilities it is easier than ever to create a YouTube Channel. You can create some fun videos and upload them directly from your phone or computer. From there you can share them on your website *(Remember that all videos uploaded directly to your website will take up the bandwidth [the companies designated internet space] and server space. Free websites have limits to both)* as well as all your social media pages. Uploading all your videos to your own YouTube channel makes a lot of sense.

Creating book trailers for each of your books is simple these days with some of the free tools available.

[That is also a paid service that I provide to my clients-https://southerndragonpublishing.com]

More readers are looking for writers on YouTube. They want to see them, hear them, and feel connected to them. Reaching out to your fans early and being consistent in your posting will garner you more followers, more quickly. The videos do not have to be long just recorded in a well-lit area, good sound quality, and, most of all, have a little bit of fun with the subject matter.

YouTube is owned and operated by Google these days. It has paid features for viewing regular movies and music too. But you can still create your very own channel, decorate it with channel art, and upload your profile picture. You can also add links to your website and up to 4 other social media pages right there in your header.

To be effective and to gain your very own channel name instead of a bunch of letters and numbers, you must do a couple of things.

Custom URL eligibility

To create a custom URL for your channel, your account needs to:

Have 100 or more subscribers

Be at least 30 days old

Have an uploaded photo as a channel icon

Have uploaded channel art

The hardest part of that is getting the 100 or more subscribers. The only way to do that is to post great short videos consistently with valuable information. Then share, share, share. It is important to comment and 'Like' your writer friends' videos and subscribe to all their channels. Again, helping each other along is always a promising idea.

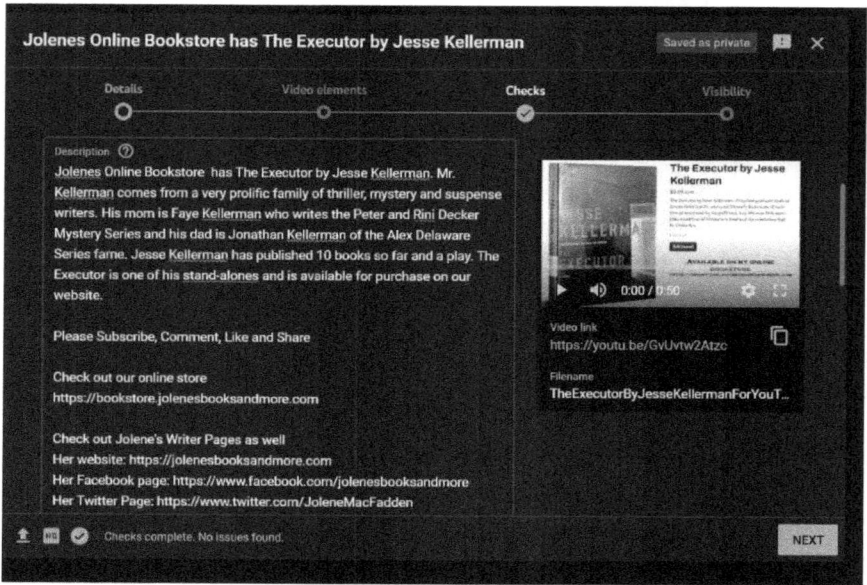

Here is a final tip for FREE about YouTube. When you type up your description of your video, please be specific and descriptive then towards the bottom you can put information about how to get in contact with you including

Your business website

Your author email address

Author public Facebook page link

Twitter Account link

Any other social media pages

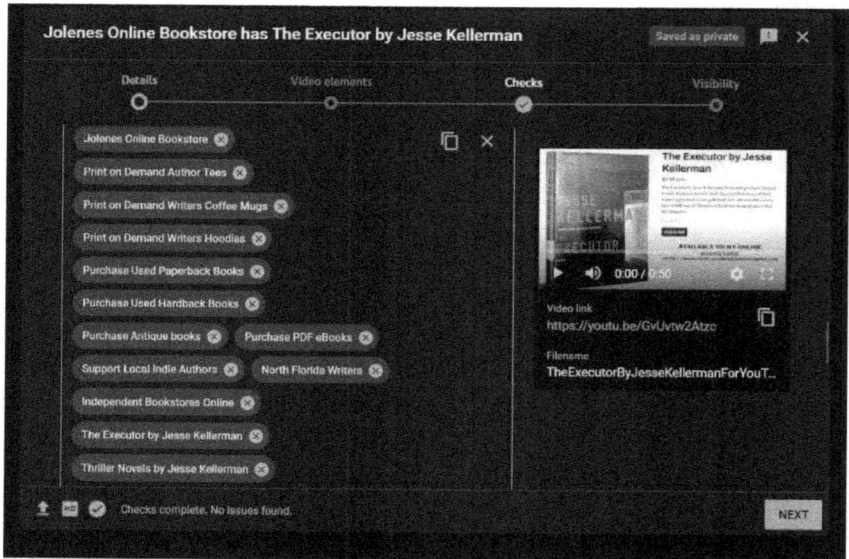

For some reason, YouTube does not like anyone putting their Amazon Profile

Page link down in the description box. Also, make sure you use descriptive tags in the 'tags' box area, not generic ones but specific to you, your writing, and that video subject.

YouTube Channel Link: _____

User Id: _____

Password: _____

Video Ideas: _____

_____

_____

_____

_____

_____

_____

_____

_____

_____

_____

_____

_____

_____

_____

_____

Quick Summary

Most of the social media platforms are pushing the 60-30 and 15 second videos in portrait format

7 to 60 Seconds called Shorts for YouTube –

Aspect Ratio 9:16 (Portrait Mode) or Size 1080x1920px (pixels)

Longer explainer videos and regular content are in landscape or Aspect Ratio 16:9 (Landscape Mode) or Size 1920x1080px.

You can specify which defaults depending on the program and equipment that you use. You will have slightly different requirements for the other social media platforms depending on the type of video, the length and where and how you are posting it. – **SEE THE BONUS PAGES**

The Hardest Part is Consistency

After all this information, we finally come to the hardest tip to carry out and maintain. YOU must create a posting schedule and stick to it. You can mark up your calendar pages, send email alerts to yourself, create an Excel Spreadsheet, or even just set alarms on your phone to remind you to do it.

Most website blog software programs, Facebook Public Pages, and even Twitter and Instagram allow you to create all your posts and schedule them as much as a month in advance. These platforms will release those posts on the days and times you show. There are some free tools out there to help coordinate your social media posting efforts so keep looking and trying new things.

I would suggest that you try to spend 15 to 30 minutes in the morning (afternoon or evening) checking all your social media accounts, retweeting, commenting, sharing others' posts and loving/wowing/ or whatever emoting you would like to let people know you have seen what they have posted. Not only do you want people to like and share your posts, but you will need to do the same for others.

CONSISTENCY, patience, and perseverance are the keys to any winning marketing campaign no matter what the product. Yours is unique to you so you need to figure out what works best for you. Contrary to that, you cannot be afraid to try new things occasionally, either. The ways to market yourself and your books are ever-changing. Being aware of what is out there, what people are talking about, and using the free seller's tools on places like Amazon you can get a better feel for what people are looking for today.

Preliminary ideas for a schedule for each of accounts you have below. Doing it a month at a time so that you have one hitting every day of the week

Facebook/Instagram: (Every Other Day) _____

_____

_____

_____

Twitter: (daily)_____

_____

_____

_____

_YouTube: (Alternate Days from Facebook) _____

_____

_____

_____

Blog: (weekly)_____

_____

_____

_____

Pinterest (daily) _____

_____

_____

_____

LinkedIn (once a week) _____

_____

_____

_____

Recap

1. Get a Website/Blog to connect to your readers.

2. Complete Your Author Profile on Amazon (and any of the other places your books are sold).

3. Create a Facebook Public Author Page.

4. Complete a Resume on LinkedIn and update it regularly.

5. Instagram, Pinterest, TikTok, Twitter and Tumblr are all free places to share your creativity in video, pictures, quotes from your books, etc.

6. Join Goodreads and fill out your author profile, attached your published books and link to your blog. Try BookBub too.

7. Write and publish your autobiography on Wikipedia.

8. Create fun and informative videos and post them on your YouTube Channel.

9. Create a Posting Schedule for your blog, social media and YouTube Channel and STICK TO IT!

10. If you are a North Florida Writer then Join the North Florida Writers Blog Tour.

North Florida Writers Blog Tour

I have talked about everything but the last one on the list. That is because it is only being offered to published North Florida authors to help bring awareness to the reading public that we have some genuinely great writers living right here in the North Florida area. Most of the counties other than Duval County are considered rural counties. I would argue that Alachua where Gainesville and the University of Florida exists, and St. Johns County with St. Augustine are urban counties. But I digress.

Rural areas do not have the resources that larger areas like Miami, Tampa, and even Orlando have in supportive Writers Groups, National Writer's Associations having local chapters, and library systems that encourage local writers to have book signings all through the year. With the crazy things happening in the world, our forward momentum in creating a video and article for each day of the week but Sunday has come to a halt.

I started the North Florida Writer's Tour to inform the public of the writers who live near them, introduce their published works, answer a few writing questions, and have the authors on tour create a short 30-second video answering one of their favorite interview questions. The authors get to choose which ones they answer.

This approach has created a wonderful diversity of information. I take the videos they send me and create a short but informative trailer for the article and the writer. They are posted and shared on all my social media pages. I email the completed posts to the writers so that they can share them on theirs.

This is not a unique thing to do, and other writers can do the same thing on their websites by writing about each other, posting, sharing, commenting, and liking each of the articles on the websites and social media pages.

Helping your fellow writers out from your writer's groups and writer's associations should be encouraged by all authors. I would say, needed, but you cannot make people do, what they do not want to do. Similarly, you cannot expect others to help you when you are unwilling to help them or even yourself.

Visit the North Florida Writers Tour website at

https://northfloridawriterstour.com

You can read some of the profiles I have already created. There is information to help you create your own videos and you can even upload your own articles if you like. Otherwise, there is a page on the site to send me your information and upload your short video. I will do my magic and post a great article about you, and your books.

Other Sites That Will Feature You and Your Books

Another place we can always include in our lists of free marketing opportunities are the following:

Writer's Podcasts

Being a guest on a podcast that is popular is a great way to get the word out for your books. Depending on what the show is about you can approach it several different ways. Make sure you understand the requirements of the podcaster, the equipment needed to participate (Most are done over the internet these days), and if you must provide any material prior to the show. This is where your publicity package in PDF format will come in handy.

There are a couple of places you can get matched up with a podcast. The one recommended the most seems to be the private Facebook group(s) below. Each offer the same possibilities of matching up guests and shows. One has more members than the other.

***Be careful that if you agree to be a guest on a show, they will not charge you for that privilege. By the same token, unless otherwise specified, you should not charge them to have you as a guest.*

Podcast Guest Collaboration Community - Find a Guest, Be a Guest

Private Facebook Group (36K members)

https://www.facebook.com/groups/podcastguestcollaboration/

This group is to connect podcasters looking for guests and those who wish to be a guest. It seems to have a waiting list to join and does not allow any promotions or fees being charged to be a guest or as a guest.

Podcast Guest Connection

(private group - 12.1K members)

https://www.facebook.com/groups/podcastguestconnection

This group is for people to network and get guests on their podcasts and possibly be guests on others shows! You should not be promoting your own podcasts episodes!

Hey! I have my own podcast now called – Jolene's Book and Writer Talk on iTunes. It is in the beginning stages, and I am always looking for guests from all areas of the writing and publishing world. Check out the link below to listen to what I have published so far and if you would like to be a guest use my Contact Form on my main website:

https://SouthernDragonPublishing.com

https://podcasts.apple.com/us/podcast/jolenes-book-and-writer-talk/id1619792506

Local Libraries

They have local writing events throughout the year, but you can volunteer for one or two of their learning sessions, lead a book club, teach younger people about the writing and publishing experience.

Local Bookstores

Not only for book signings but offering writing classes, leading book discussions, reading to the kids, and more.

Local Schools

Introduce the kids to the world of writing, share your publishing journey, and on Career Day explain how to become a published author.

Each of the above places will generally have their own websites and social media pages to promote all events that you agree to participate in or run yourself. It is a great way to cross promote each other.

Author's Websites that will feature other writers on their blog, newsletter, social media platforms and/or podcast.

You, in turn, should agree to reciprocate in a like manner on your website and/or social media pages.

_____

_____

_____

_____

_____

_____

_____

_____

_____

_____

_____

_____

_____

_____

_____

_____

_____

_____

SOUTHERN DRAGON
Publishing

Thank you so much and I hope you enjoyed this book. If you found it useful, please comment on whichever selling site you found it on.

You can email me at Jolene.macfadden@jolenesbooksandmore.com

 or visit one of my websites and use the contact forms.

https://jolenesbooksandmore.com

For all those North Florida published writers out there, who haven't gotten with me to create your FREE featured article and promo video check out the website.

https://NorthFloridaWritersTour.com

For those who would like to know more about all of my paid services please visit my Southern Dragon Publishing website.

https://southerndragonpublishing.com

BONUS PAGES

Set Up Your Social Media

Choose your color schemes (2 or 3) and your preferred fonts. The User Name Should be Your Writer Name (pen name). Having the same name for each platform helps keep things organized.

Facebook Profile Pic 170 x 170px
Facebook Cover Photo 851 x 315px

Instagram Profile Pic 320 x 320px

LinkedIn Profile Pic 400 x 400px
LinkedIn Cover Photo 1128 x 191px

Twitter Profile Pic 400 x 400px
Twitter Header Cover 1500 x 500px

YouTube Profile Pic 800 x 800px
YouTube Channel Art 2048 x 1152px

9:16 - 1080x1920px

Instagram Reels
TikTok Videos
Facebook Stories and Reels
YouTube Shorts
Pinterest Pins
LinkedInVideo Ads

1:1 - 1080x1080

Pinterest
Instagram
Twitter

YouTube Videoss
LinkedInVideo Ads

16:9 - 1920x1080px

Scheduling Versus Posting

Facebook and Instagram Posts you can schedule on your Business Page - pictures and videos

You can create articles in LinkedIn and save as drafts

TikTok - Schedule Videos but not create them

You can schedule Regular Pinterest Pins but Not Idea Pins

Can Schedule Tweets up to a month in advance on Twitter

Can Save Pinterest Idea Pins as drafts and post each day

You can only create Facebook reels on cellphone - save them as drafts to post later

You can create movies in TilkTok and save them as drafts

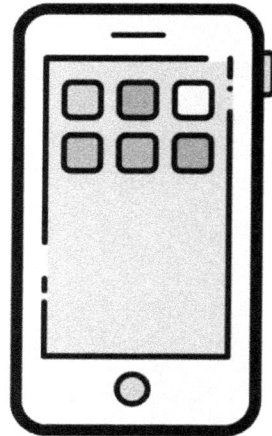

Another bonus we have available for anyone who purchases this book is a Monthly Calendar to fill with ideas based on the special days coming up, to add any writers' meetings you have and want to remember as well as any upcoming book festivals. All of these are great topics for posts on your website and social media pages.

In addition, I have added a weekly social media Posting Schedule Page, A Weekly Writer's Calendar and finally one Book review form to encourage you to read a book and review it at least once a week. So that's one generic monthly calendar and copies of each of the forms mentioned above and that I have pictured in this book to get you started. You can make your own forms with what works for you. These are just to get you started.

Some Posting Ideas to Ponder

A great social media posting idea is a post about whichever 'National Day of---' is that day. My favorite is anything to do with chocolate or coffee. You know, 'June 7th - National Chocolate Ice Cream Day'. Now create some video of you eating chocolate ice cream while typing on your computer.

You can always create posts or short videos of you writing, talking about your writing process, how you find time to write, what music you play while you are writing, etc.

A good practice for upcoming interviews on other writer's websites and podcasts, even possible local news stations and the like is to have a list of the top 100 frequently asked questions. You get to prepare ahead of time your answers and create little posts, pictures with quotes of the questions and your answers to add to your social media feed.

While you are out for a walk anywhere and for whatever reason, take some time to snap some pictures of your surroundings. Even take a little video or two of birds flying, fish jumping, water falling down a mountain, you know whatever is happening around you at that time and use it with music and text-overs for your posts.

SAMPLE QUESTIONS

Listing from one of the homeschooling websites

When did you first realize you wanted to be a writer?

How long does it take you to author a book?

What is your work schedule like when you're writing?

What would you say is your interesting writing quirk?

How do books get published?

Where do you get your information or ideas for your books?

When did you write your first book and how old were you?

What do you like to do when you're not writing?

What does your family think of your writing?

What was one of the most surprising things you learned in creating your books?

How many books have you written? Which is your favorite?

Do you have any suggestions to help me become a better writer? If so, what are they?

Do you hear from your readers much? What kinds of things do they say?

Do you like to create books for adults?

What do you think makes a delightful story?

As a child, what did you want to do when you grew up?

Imaginative and Fun Author Interview Questions
From: https://TheAuthority.pub

How about 100 more to keep you going. The great thing is you can pick and choose the ones you like the best then just cross them off as you answer them either in a blog post, on a social media picture post (meme) or creating a short video of your answer. These will give you lots of great material to get you through the coming weeks of posts.

Best Author Interview Questions

1. At what point do you think someone should call themselves a writer?

2. What difference do you see between a writer and an author?

3. Have you ever considered writing under a pseudonym, and why or why not?

4. What do the words 'writer's block' mean to you?

5. How do you process and deal with negative book reviews?

6. Are there therapeutic benefits to modeling a character after someone you know?

7. What is the most difficult part of your writing process?

8. How long have you been writing or when did you start?

9. What advice would you give to a writer working on his/her first book?

10. What, to you, are the most essential elements of skillful writing?

11. What comes first for you — the plot or the characters — and why?

12. How do you develop your plot and characters?

13. When did you first call yourself a writer?

14. How do you use social media as an author?

15. What's your favorite and least favorite part of publishing?

Author Interview Questions about Their Book

1. How many books have you written, and which is your favorite?

2. What part of the book did you have the hardest time writing?

3. What part of the book was the most fun to write?

4. Which of the characters do you relate to the most and why?

5. If you're planning a sequel, can you share a tiny bit about your plans for it?

6. What is a significant way your book has changed since the first draft?

7. What perspectives or beliefs have you challenged with this work?

8. What inspired the idea for your book?

9. How would you describe your book's ideal reader?

10. How much research did you need to do for your book?

11. How important was professional editing to your book's development?

12. What was your hardest scene to write, and why?

13. What characters in your book are most like you or people you know?

14. How long did it take you to write this book?

15. How did you come up with the title for your book?

16. Would you and your main character get along?

17. If you could meet your characters, what would you say to them?

Fun Author Interview Questions

1. What is your writing process like? Are you more of a plotter or a pantser?

2. What do you need in your writing space to help you stay focused?

3. If you were to write a spin-off about a side character, which would you pick?

4. If you could spend a day with another popular author, whom would you choose?

5. What is your schedule like when you're writing a book?

6. Have you ever traveled as research for your book?

7. What's your favorite writing snack or drink?

8. How do you celebrate when you finish your book?

9. What do you think of NaNoWriMo? Worth it?

10. What is your kryptonite as a writer?

11. What risks have you taken with your writing that have paid off?

12. When was the last time you Googled yourself and what did you find?

13. Which of your characters are most likely to be an activist, and what kind?

14. Do you play music while you write — and, if so, what's your favorite?

15. Have pets ever gotten in the way of your writing?

16. If your story was made into a movie, which actors would play your characters?

17. Have you ever killed off a character your readers loved?

Questions about Writing

1. What is the most valuable piece of advice you've been given about writing?

2. What do you think is the best way to improve writing skills?

3. What advice would you give to help others create plotlines?

4. What has helped or hindered you most when writing a book?

5. Does writing energize or exhaust you? Or both?

6. What is the best money you've ever spent with regard to your writing?

7. What are common traps for new authors?

8. How many hours a day do you write?

9. What are your favorite blogs or websites for writers?

10. At what time of the day do you do most of your writing?

11. What's your writing software of choice and why?

12. How do you come up with character names for your stories?

13. Do you take part in writing challenges on social media? Do you recommend any?

14. When you're writing an emotional or difficult scene, how do you set the mood?

15. Who do you trust for objective and constructive criticism of your work?

16. What are the essential characteristics of a hero you can root for?

17. What do you do to get inside your character's heads?

Questions to Ask Authors about Other Books and Authors

1. What books do you enjoy reading?

2. Are there any books or authors that inspired you to become a writer?

3. What books helped you the most when you were writing your (first) book?

4. What books did you grow up reading?

5. What authors did you dislike at first but then develop an appreciation for?

6. Name an underappreciated novel that you love and tell us a little bit about it.

7. Has any hugely popular novel left you thinking you could write it better?

8. Have you ever tried to write a novel for a genre you rarely or never read?

9. What book (or books) are you currently reading?

10. If you could be mentored by a famous author, who would it be?

11. When you are reading for enjoyment, do you prefer eBooks, printed books, or audiobooks most of the time?

12. What are your favorite series or series authors?

13. Have you listened to any audiobooks? Which did you enjoy the most?

14. If you could be a character in one of your favorite books, who would you be?

15. What author in your genre do you most admire, and why?

16. Have you used an app to borrow eBooks or audiobooks from the library? If so, which one(s)?

17. What books have you read more than once in your life?

Personal Questions for Authors

1. Has writing and publishing a book changed the way you see yourself?

2. Is there a particular genre you would love to write but only under a pseudonym?

3. Do you see writing as a kind of spiritual or therapeutic practice?

4. As a writer, what would you choose as your spirit animal?

5. What spiritual or therapeutic practices help you get into the right headspace?

6. At what stage (or stages) of your life have you done most of your writing?

7. What's the trickiest thing about writing characters of the opposite gender?

8. What do the words 'literary success' mean to you? How do you picture it?

9. If you didn't write for a living, what would you probably do for work?

10. Would anyone in your family disapprove of anything you've written?

11. Does anyone in your family read your books?

12. Who has been the biggest supporter of your writing?

13. Do you have other writers in the family?

14. If you could invite any three people for dinner, who would you invite?

15. Would you share something about yourself that your readers don't know (yet)?

16. If you had to describe yourself in just three words, what would those be?

17. If you had the power to cure a disease of your choosing, what would it be?

SAMPLE FORMS AVAILABLE FOR DOWNLOAD

MONTHLY CALENDAR

Year:____

Sunday	Monday	Tuesday	Wednesday	Thursday	Friday	Saturday

Special Days:

WEEKLY POSTING PLANNER

Digital Marketing can Help You sell More!

For the Week of: _____

Platform(s)

Monday					
Tuesday					
Wednesday					
Thursday					
Friday					
Weekend					
End of the Week Stats					

Ideas for next week

Include a mixture of Reels, Stories, Regular Posts and Live(s)

AUTHOR'S WEEKLY TO-DO'S

Dates: _____

Word Count Goal: _____

Writing Tasks To Do   *Business Tasks To Do*

	Monday	
	Tuesday	
	Wednesday	
	Thursday	
	Friday	
	Saturday	
	Sunday	

Time Off Is Important Too   Final Word Count

BOOK REVIEW
★★★★★
Overall Rating

Book Title: _____   Genre: _____
Author: _____   Date Published: _____
Date Read: _____   Review Posted: _____
(Goodreads, Amazon, BookBub, Social Media, Other)

QUICK SYNOPSIS:
_____
_____
_____
_____

MAIN CHARACTERS(S)   THEME(S) - TROPE(S) - SETTING
_____   _____
_____   _____
_____   _____
_____   _____

WHAT DID YOU LIKE ABOUT THE STORY AND CHARACTERS
_____
_____
_____
_____

WHAT DID YOU THINK NEEDED IMPROVEMENT IN THE STORY AND/OR CHARACTERS
_____
_____
_____
_____

GOOD LUCK

CONTACT THE AUTHOR

I wanted to thank my family for their patience while I have been trying to get this book updated and republished.

If you would Subscribe, Like, Share and Comment on my social media pages and subscribe to my website newsletters, I would be forever grateful. Of course, I would like to do the same for you so don't be shy. Send me a direct message on any one of these accounts and let's support each other's efforts to become a successful independently published author!

https://jolenesbooksandmore.com

https://southerndragonpublishing.com

https://NorthFloridaWritersTour.com

https://twitter.com/JoleneMacFadden

https://www.facebook.com/jolenesbooksandmore

https://www.instagram.com/jolenesbooksandmore/

https://www.linkedin.com/in/jolene-macfadden-kowalchuk-29a55825/

Jolene's YouTube Channel Has Lots of Great Videos

https://www.youtube.com/c/jolenesbooksandmore2669/

Visit my bookstore and buy some books or merchandise:

https://bookstore.jolenesbooksandmore.com/

Join in on the Fun at Jolene's Patreon – subscribe and get valuable information, discussions, videos and more:

https://www.patreon.com/JolenesBooksAndMore

I am even creating a Jolene's Writers and Book Talk Podcast -HINT – HINT – I would like to interview you about your writing process and your books. Heck, select some of those interview questions from earlier and let's get you on the air and shared everywhere!

https://podcasts.apple.com/us/podcast/jolenes-book-and-writer-talk/id1619792506

I look forward to hearing from you soon.

ABOUT THE AUTHOR

Jolene MacFadden is a published author (5 non-fiction books), blog writer, website designer, digital marketer and virtual assistant to other authors and small businesses. After working in the medical office field for about 20 years she retired to travel around the state with her mom in an old RV. Sadly, they had to stop as the RV blew an engine and her mom's health was steadily declining. Now she has her own online bookstore and crafts store which she hopes to expand to have a real "Sticks N' Bricks" place of her very own. Jolene freelances as a publishing and writers assistant, social media coordinator (digital marketer) and website designer for authors through her publishing business website, Southern Dragon Publishing. She also offers her virtual assistant, digital marketing and website design services to local small businesses through her Jolene's Web Designs. Jolene has ghostwritten a couple of non-fiction books over the last year with more to come. Now she hopes to have her first fiction book published before the end of 2022.

The super-secret page to download all of the forms and more discussed here in this book:

https://southerndragonpublishing.com/readers-bonus-page/

This is a password protected page:

SDPbookbranding123

BOOKS BY THE AUTHOR

Insider's Guide to Campground Hosting in Florida Parks

Insider's Guide To Campground Hosting In Florida Parks is for those who would like to spend some time in some of our great county, state and national parks and are willing to work in exchange for a campsite. We give the newcomer the skinny on how to apply, where to find the best places to volunteer and information on the benefits of helping our hard-working park staff.

There over 100 Florida State Parks, various Florida State Forests, Florida County Parks and a few National Parks and Forests that will give you a FREE campsite in exchange for volunteering. We will even explore some opportunities with our Florida Fish and Wildlife, Regional Water Management Districts, as well as other government agencies that can offer you a campsite with water, electric, possibly sewer on-site and more for a little bit of your time and sweat that will help keep our beautiful resources cleaned, maintained, and repaired for all to enjoy.

We have visited and/or worked in quite a few of the Florida State Parks over the last 2 years and have gathered some great information for our fellow RVers. This is a great guide for people wanting to travel in a RV in the future, those just getting started as well as those who have been traveling, living, and working in their RVs for a while. Being a volunteer in public parks and land areas, whether they are county, state or national, helps keep them up and running, gives you a sense of accomplishment, and it even can give you a place to stay in a beautiful setting. Of course, you will be meeting and working with some interesting people, learning new skills, and enjoying the great outdoors.

Top Ten Ways to Market Your Book for Free

Quick and easy information for writers who want to know how to market their book on a very tight budget. Tips and tricks gathered from years of working with writers, publishing my own books, and as a reader wanting to find more on the writers I want to read

Suwannee River State Park: Activities, Games, Record Your Adventures and Share

Suwannee River State Park is a full-service recreational area located in Live Oak, Suwannee County Florida. We are in the process of creating workbooks for kids to enjoy and fill out while visiting any of our Florida public land areas that will include puzzles, coloring pages, activity sheets and journal pages. We hope that you and your child(ren) will cherish these memories and record them using the journal pages inside. You can take pictures and paste them in and/or your child can draw them. We have created the puzzle pages from information from this park's official website so visit it, read about it's history, activities and other great information. Supporting our great public land areas by visiting them, recording your adventures and sharing them with friends and family is a great way to keep them open and available to the general public for years to come.

Anastasia State Park: Activities, Games, Record Your Adventures and Share

This cute workbook is for kids and their parents to have some extra fun while visiting or camping inside Anastasia State Park in St. Augustine, Florida. It is a lovely place to spend the day or even a couple of days. They have lots of activities all through the year and close to some great attractions and other historical sites. Best of all it is right on the Atlantic Ocean. The activity book has pictures to color, puzzles to do and a place to record your adventures and your children draw pictures of what they see and experience. Printing out some of your pictures while you are then and taping them inside the book will create a unique keepsake for you and your children for years to come.

Workbooks for Fun

Create Your Own Limericks Workbook

We all enjoy poetry in one form or another. One of the simplest forms has to be the Limerick. All you need is some words that sound close to the same, a little patience, and tell a very short story in the form of AABBA We have included some samples from various sources to serve as examples. You can just change them up with your own words or create your own. This workbook has plenty of space for you to write down your inspirations as they happen. Once you fill it up you can publish your very own book of poetry.

Passed Down Recipes: From all the Past Generations of Men and Women in Our Family to Future Generations to Come

Gathering Recipes is a time-honored tradition in almost all families. There is always one or two people who have kept the family recipes that have been recorded on scraps of paper, clipped out of newspapers or written down on index cards. They contain the notes from family chefs throughout the ages of what worked and what didn't. We usually only get to eat them during holidays, birthdays, and funerals.

Use this book to help organize all those recipes into sections, share some memories about the people who made them through the years and draw or paste in some pictures too. It will make a great keepsake for your future generations to use and enjoy.

Book Review Workbook: Practice Makes Perfect - Become a Professional

Writing a book review for your favorite writer is a great way to show your support. Yes, purchasing their books, reading their books, and sharing their books with your friends and family is a great way to start. These days it is important to also write, helpful, informative, and honest reviews of each story as you read them. Sharing those reviews on Amazon, Goodreads, and other book sites is a great service you can do for all published authors.

After filling out one of our workbooks you might even be able to start a whole new career doing book reviews on your own YouTube and/or TikTok channels. Our review pages have some general questions to guide your reviews and plenty of space to write out your thoughts. Typing these in will be easier once you have written them down. Also, it a great place to start on a video script. Good Luck and we hope to be seeing your reviews and videos soon.